HISTORIC PHOTOS OF
COLORADO MINING

TEXT AND CAPTIONS BY ED RAINES

TURNER
PUBLISHING COMPANY

When this photograph was taken of the Gregory gold vein, almost 40 years had passed since its 1859 discovery. A railroad served the district; various structures had been built and torn down; and the richest, near-surface portion of the vein had been completely depleted, or mined out, leaving the open cut. The Gregory-Bobtail vein system produced about $20,000,000 (gold values when mined) between 1859 and 1959.

HISTORIC PHOTOS OF
COLORADO MINING

Turner Publishing Company
www.turnerpublishing.com

Historic Photos of Colorado Mining

Copyright © 2009 Turner Publishing Company

Library of Congress Control Number: 2009922625

ISBN: 978-1-59652-535-1

Printed in the United States of America

ISBN 978-1-68442-089-6 (hc)

Contents

The "Ice Palace" was a regular sightseeing stop on the Argentine Central Railroad's McClellan Mountain route. Here, a tourist examines an abandoned mine tunnel encrusted with a beautiful growth of sublimated ice crystals.

Acknowledgments

This volume, *Historic Photos of Colorado Mining,* is the result of the cooperation and efforts of many individuals, organizations, and corporations. It is with great thanks that we acknowledge the valuable contribution of the following for their generous support:

The Colorado Historical Society
Colorado School of Mines Arthur Lakes Library
Colorado School of Mines Geology Museum
The Denver Public Library
The Library of Congress

We would also like to thank Silvia Pettem for her valuable contributions and assistance in making this work possible.

With the exception of touching up imperfections that have accrued with the passage of time and cropping where necessary, no changes have been made. The focus and clarity of many images is limited to the technology and the ability of the photographer at the time they were taken.

PREFACE

My parents, both Texans, met in Estes Park, Colorado, in the late 1920s. Throughout the rest of their lives, they came to Colorado almost every year. After my fourth birthday, they brought me with them. That's what Texans did in the pre-air-conditioner summer days of 95- to 105-degree heat: they got in their cars and drove two days in that heat so they could relax in the mountains for a few weeks, before they had to climb back in their cars and drive through that heat again. It was worth it, a thousand times over.

When we came to Colorado, we always brought along Muriel Sibell Wolle's *Stampede to Timberline.* As we drove from one old mining town to another, Mom would read aloud Mrs. Wolle's descriptions and stories of the place where we were headed. By the time that I was 18, I had visited 80 percent of those old mining areas that Mrs. Wolle described. And I had started a hobby.

Mom and Dad were avid fishermen. At the age of nine or ten, while standing alongside some mountain stream, I decided that I was not. I was bored stiff and I wasn't going to fish anymore. So I picked up some rocks and began to work on my curve ball. I threw rocks at all the surrounding trees and retired by the side. Before the next inning could start, my mind had shifted scenes, and I was lobbing hand grenades at some imagined enemy. I discovered that if I threw my grenades into the stream, the ensuing splash would approximate the explosion that my imagination required. Mom's words brought me back from the Battle of the Bulge. "Ed, you are ruining the fishing. Why don't you look around and see how many different kinds of rocks you can find? Make a collection and we can identify them." So I did. And I still collect rocks. I don't fish.

Every place that we went in Colorado had rocks, and in lots of those places people have mined them. I knew, because Mrs. Wolle had told me so. So if we went someplace where silver was mined, I tried to collect silver ore, and the same with gold, zinc, lead, tungsten, et cetera. I soon learned that it was difficult for a beginner to find good rocks.

Then in my sixteenth summer, I was fortunate enough to meet Stan Noga, a prospector and the owner of a small rock shop near Buena Vista, Colorado. Stan took me collecting, explained that mineralogy was a science, and introduced me to the literature of geology, mineralogy, and mining. I began to read whatever I could find (and understand) that would tell me where to look. I was hooked.

Later, when I studied geology in college, a graduate student told me that there was a day that every geologist could look back and say that on that day, he became a geologist. He was right; mine came when I discovered that, from cover to cover, I could read and understand Samuel Franklin Emmons's U.S. Geological Survey Professional Paper 148, "The Geology of the Leadville Mining District, Colorado." I also read the USGS papers, bulletins, monographs, and annual reports on Cripple Creek, Breckenridge, Central City, Georgetown, Lake City, Creede, Silverton, Telluride, Bonanza, and lots of others. My studies, work, and collecting have taken me to lots of other places in the world, but I always seem to come back to Colorado. In fact, I moved here in the late 1980s and married Silvia Pettem, who shares my interest in mining history.

There really are only two basic industries in this world—mining and agriculture. A wonderful Art-Deco bas relief over the front door of the Boulder County Courthouse illustrates that basic fact to all who come to do business with the county. For years, those two industries were, economically, Colorado's most important, and it was mining that paved the way to statehood. I have spent most of my life studying that industry.

The story of Colorado mining is intriguing. As the consulting Collections Manager at the Colorado School of Mines Geology Museum, I am sometimes asked what book to read about Colorado mining. I always say *Stampede to Timberline,* a copy of which sits on the shelf by my desk. Mrs. Wolle was an artist, as well as a wonderful storyteller, and her book is filled with her drawings. Pictures can speak louder than words, and this book, *Historic Photos of Colorado Mining,* is meant to be an introduction to Colorado mining history told through photographs, each of them with identifying captions. I couldn't cover everything—the subject is truly vast—but I tried to provide a solid introduction to a wonderful story. I hope you will enjoy that story as much as I do.

—*Ed Raines*

NEDERLAND

CARIBOU

SILVER PLUME

CENTRAL CITY
& Black Hawk, Apex

GILMAN

CLIMAX

LEADVILLE
& California
Gulch, Oro

IDAHO SPRINGS

GEORGETOWN

COLORADO

PLATEAU

ASPEN

BRECKENRIDGE

FAIRPLAY

CRIPPLE CREEK
& Victor, Independence, etc.

OURAY
& Sneffels

RED MOUNTAIN

TELLURIDE

LAKE CITY

ORIENT

WAGON WHEEL GAP

CREEDE

DURANGO

SILVERTON
& Gladstone, Eureka,
Animas Forks

Moffat | Routt | Jackson | Larimer | Weld | Logan | Sedgwick

Phillips

Grand | Boulder | Morgan | Washington | Yuma

Rio Blanco

Eagle | Gilpin | Jeffer-son | Adams

Garfield | Summit | Clear Creek | Den-ver

Mesa | Pitkin | Lake | Chaffee | Douglas | Arapahoe | Kit Carson

Delta | Park | Teller | Elbert

Montrose | El Paso | Lincoln | Cheyenne

Crowley

Ouray | Fremont | Custer | Kiowa

San Miguel | Gunnison | Hinsdale

Dolores | San Juan | Saguache | Huerfano | Pueblo

Mineral | Rio Grande | Alamosa | Otero | Bent | Prowers

Montezuma | La Plata | Archuleta | Conejos | Costilla | Las Animas | Baca

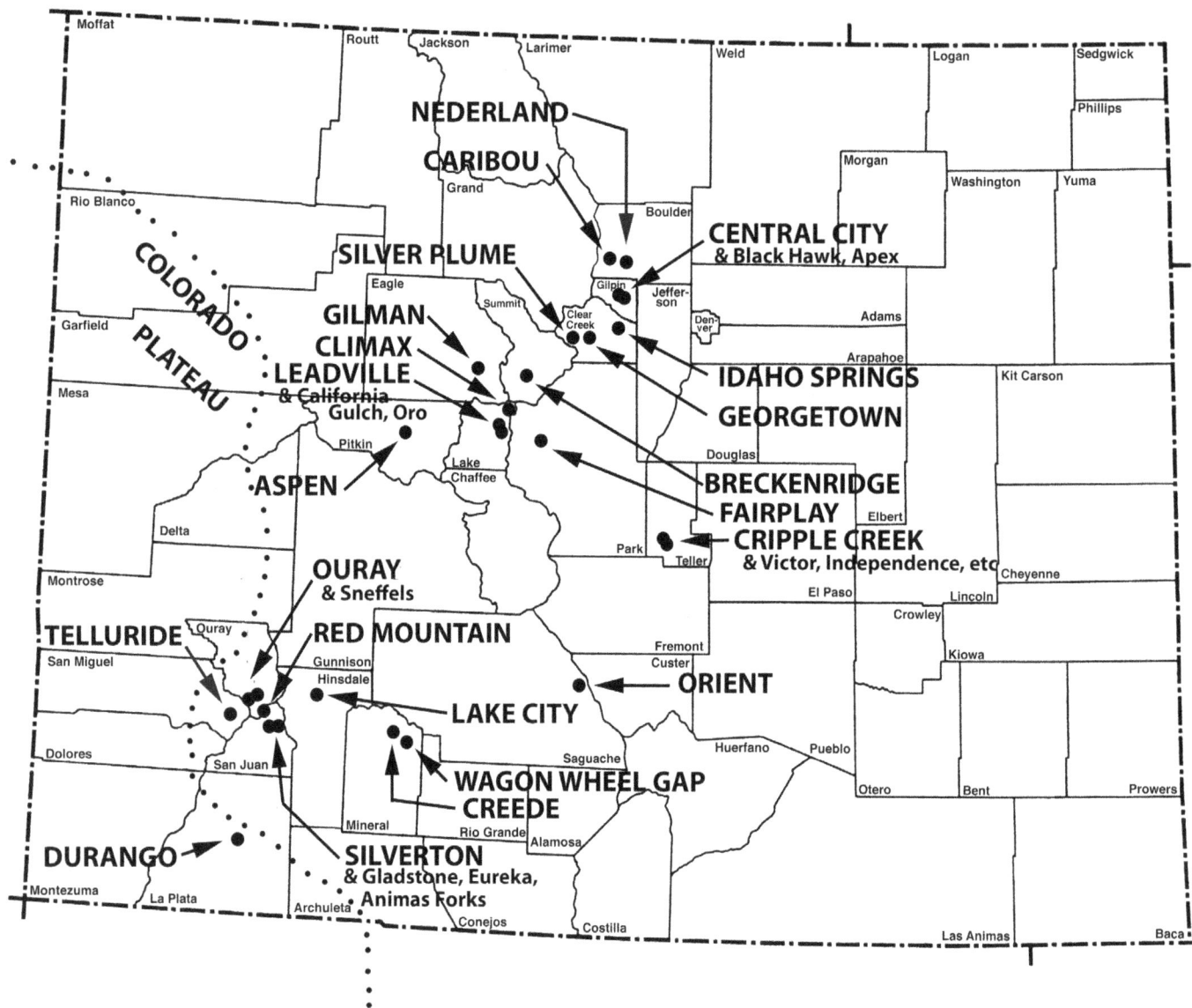

This map of the state of Colorado shows mining towns, districts, and areas pictured and discussed in this book.

The Gold Rush and the Initial Mining Effort

(1859–1895)

In 1850, a party of California-bound prospectors discovered placer gold near present-day Denver. During the summer of 1858, Green Russell, along with several others of those same forty-niners, returned to the Colorado Rockies to further investigate the potential of the area. Gold discoveries were made along the foothills of the Front Range near present-day Idaho Springs, Gold Hill, and Central City. During the next several years, 100,000 people started for Colorado and half completed the journey, but 25,000 soon gave up and went back home. The 25,000 who stayed built a state.

Soon prospectors found the source of the placer gold in veins cutting through the mountains above the stream valleys. Lode claims were staked, and underground hardrock mining soon surpassed the placers in gold production. Local prospectors and miners in each region established mining districts and wrote rules and regulations to specify standard operating procedures concerning claim size, location, recording procedures, and water rights. Many of these districts and their rules are still recognized by government today.

There were only so many veins on a given mountain, so a district would eventually become claimed-up. Late arrivers had to move on and prospect unexplored areas. New discoveries followed old as more and more prospectors fanned out to look for gold. The towns of Dumont, Lawson, Empire, Georgetown, Silver Plume, Geneva, Montezuma, Breckenridge, Ward, Tarryall, Fairplay, Alma, Buckskin Joe, Granite, and California Gulch all resulted from the ever-widening search.

In each of the gold camps, the mining landscape changed over time. As placer deposits played out, miners sank shafts and drove adits into the veins. Sluices, dams, and gravel piles along the streams gave way to headframes, boiler plants, shaft houses, and mine dumps that perched on the mountainsides. As miners dug deeper and discovered more complex ores, mills were built to process the ores. Freighters drove multiple teams of horses or mules that pulled and pushed wagons loaded with heavy equipment up narrow, steep mountain roads. Mining engineers assembled sophisticated machinery to allow armies of miners to dig deeper into the earth. Metallurgists processed and concentrated the ore. And then the freighters returned to fill their wagons and haul the concentrates to smelters. Put another way, the engines and machinery of the Industrial Revolution had come to the Colorado Mountains.

Actual photos of the journey to the goldfields are very scarce because the fifty-niners didn't own cameras. These early "Pikes Peakers" were mostly farmers and townspeople from communities scattered along the Missouri River system. The Panic of 1857 had brought hard times, and a journey of 600 miles to what could be a new beginning seemed like a logical thing to do. As historian Elliott West once said, "It wasn't a case of 'why go?' It was more like, 'why not?'"

DENVER CITY, K. T.—SEE PAGE 182.

Only a few crude log structures had been erected in Denver when *Frank Leslie's Illustrated Newspaper* published this engraving. K. T. stands for Kansas Territory—both statehood for Kansas and an official name for Colorado were still in future. Denver was named for James W. Denver, the governor of the Kansas Territory.

PIKE'S PEAK—OUR CAMP IN AURARIA, K. T.

Auraria was the name of the community that grew up on the south side of Cherry Creek; Denver was located on the north side. In late 1858 and early 1859, Auraria was a tent city occupied by disappointed gold-seekers—the meager amount of gold found in the sands of Cherry Creek was not sufficient to support a sustained mining effort.

A lone prospector pans for gold near the Jackson discovery site. On January 7, 1859, near present-day Idaho Springs, George A. Jackson discovered placer gold in a sand and gravel bar along Chicago Creek at its confluence with Clear Creek. Jackson kept his discovery a secret until May, when he returned to the site with a party of prospectors.

In 1861, the course of the Gregory vein was clearly marked by a series of mine structures. John Gregory's prospecting journey is truly underappreciated today. He sampled and panned his way down the Colorado Front Range from Fort Laramie, and correctly selected Clear Creek (known then as the Vasquez Fork) as a major source of gold in the South Platte River. He then followed small amounts of gold upstream, (correctly) turning north at the forks, and (again, correctly) turning west into Gregory Gulch, where he discovered a gold lode at the vein that now bears his name.

By 1879, more structures had been built along the Gregory vein. Central City's original mining rules prescribed a meager 100-foot-long claim. Several companies bought up many of these short claims in order to assemble an overall length that could be worked on a profitable basis.

GREGORY GOLD DIGGINGS, COLORADO, MAY, 1859. Page 181.

Prospectors followed the Gregory party and established a "Diggings" within a matter of days. Shown in this engraving are several sluice boxes that share the area with tents and cabins of a new community. Civilization of a sort has been initiated with the arrival of a woman in camp.

PIKE'S PEAK---PARTY OF MINERS GOING ON A PROSPECTING TOUR.

Following the Jackson and Gregory discoveries, prospectors took to the Front Range hills, mountains, and beyond. Within a few years, they discovered many mineral deposits, established numerous mining districts, and founded new mining camps—some to be abandoned and others to continue on as mountain towns.

The Denver Public Library designates this image "the earliest-known photograph of Central City." Log and rough board structures are strung along Gregory Gulch, and piles of excavated rock dot the hills. Mining has come to Colorado Territory.

Sluices (or flumes) are shown here entrenched in a gulch at Central City in the 1860s. Several miners and their wives pose for the camera along the trench edge.

In the 1860s, prospectors-turned-miners pose while working a sluice in Gregory Gulch, Central City District.

A gold pan was indispensable to a prospector. An experienced panner can recover approximately 80 percent of the minerals that are heavier than quartz. At the annual World Panning Championship held in Sacramento, it is common for contestants to recover 20 pinhead-sized grains of gold from a ton of crushed rock in less than 30 minutes.

Miners work a large ground sluice in this view of Russell Gulch, Central City District. The blurred area in the background is the spray of water from a hydraulic nozzle washing the gravel into a sluice, which would trap the gold.

Miners periodically shut down their sluices in order to perform a cleanup. In this photo they have blocked the water flow to the sluice box. They will now drizzle a little mercury into the box to mix with the gold to form amalgam, and then scrape and scoop the mixture out with a small hand shovel. The amalgam will be heated in a retort (a round glass vessel with a long neck) and the mercury boiled off and recovered as a distillate. The gold left behind is known as a "sponge," a reference to its appearance, not to its squeezability.

A few brick and stone buildings stand out in a sea of rough board structures in this view of Central City from before the fire of 1874.

In a scene typical of the industrialized, mining West, a train runs across the hill above crowded Packard Gulch, a tributary of Gregory Gulch. The principal structure is the shaft house and surface plant of the Gregory-Buell Consolidated Gold Mining and Milling Company.

As miners dug deeper in the Central City mines, the ores became more complex, and extracting gold proved more difficult. To surmount the obstacle, Nathaniel Hill, a Brown University chemist, traveled to Swansea, Wales, to study the Welsh reverberatory furnace. Hill's tests on Colorado ores were successful, and in early 1867, the Boston and Colorado Smelter was "blown-in" (began smelting). The reverberatory furnace saved the Colorado mining industry, and Hill went on to become a U.S. senator.

Almost all of Colorado's mines were discovered by prospectors. In this scene from 1881, a group of prospectors, with their faithful burro, have established their camp at timberline.

After doing his stream-side laundry, a prospector's hands might be numb for 20 to 30 minutes!

Hydraulic mining was king in the Fairplay placers.

California Gulch was the biggest boom of 1860. Several thousand people settled in for one of the most hectic summers that Colorado would see during the next two decades. They recovered between $3,000,000 and $5,000,000 in gold, but who was counting?

Placer gold deposits are formed by the action of running water, often from ancient streams. Over time, however, a stream channel can migrate, leaving gold-bearing sands and gravels "high and dry." In order to have enough water to operate a large placer mining operation, miners had to bring the water to the gold-bearing gravels through a series of ditches and flumes. In this photo, a crew is grading an incline to support a flume.

Water was brought to the Keystone placer mine along a graded ditch lined with a wooden trough.

A miner shoots a powerful stream of water through a hydraulic giant monitor (nozzle) over gold-bearing stream gravels. Water from the monitor is flowing toward the two wing boards in the background, which channel the water and gravels into a sluice box.

At the Alma placer in Park County, two monitors shoot water at the gravel beds. A waterfall is adding even more water to help move the gravels along toward the sluice box, visible at the lower left.

Breckenridge was home to the largest placer mining operations in the state. In this photograph, two monitors are seen at work on Farncomb Hill. Some of Colorado's finest gold specimens were mined at this location.

On July 23, 1887, Tom Groves and Harry Lytton, lessees operating on the Gold Flake vein, mined more than 243 ounces from a single pocket in the vein. The largest gold specimen weighed 160 ounces (13.3 troy pounds) when first mined, but two pieces broke off, leaving a single mass of more than 136 ounces. The specimen as shown in this photo was featured in the Colorado Mining exhibit at the 1893 World's Columbian Exposition in Chicago. Groves displayed so much pride in the specimen that it became known as Tom's Baby. Through the years another piece broke off, so as now displayed at the Denver Museum of Nature and Science, it weighs 103 troy ounces.

John F. Campion, a well-known Leadville mining entrepreneur, consolidated most of the Farncomb Hill claims into the Wapiti Mining Company, with its office on Farncomb Hill in the Breckenridge District. This photo shows a portion of the interior of the company's office. In addition to the room's furniture and several balances, the door to a large, decorated safe can be seen at the extreme left.

Dredge boats were used with good results at Breckenridge for nearly half a century. This view shows the "business end" of the dredge: a conveyor belt of steel buckets to dig out the gold-bearing gravels of a deeply buried placer deposit. An assembly of screens and sluices traps the gold inside the body of the dredge, and the waste rock is discarded over another conveyor belt at the stern of the boat.

After the gold is extracted, the stacker (a large conveyor belt at the stern) dumps the coarse gravels into piles at the edge of the pond. The boat gradually digs its way forward, leaving a distinctive series of gravel piles as a rock wake that is still visible today.

The route that the dredges followed was not left to chance. Prospectors used a drill rig to test old, buried stream channels, and the dredges followed a path selected by a careful sampling program, in which the test results were plotted on a map.

Silver, Too!

(1867–1895)

There is more than gold in many of Colorado's mineral deposits. During the mid-1860s, assays from samples taken near Georgetown and Silver Plume indicated economic quantities of silver. Extracting silver from ore was complicated, but by the late 1860s, Nathaniel Hill's Boston and Colorado Smelting Company had mastered the process. In 1869, silver was discovered at Caribou in Boulder County. The Georgetown–Silver Plume and Caribou area produced most of the state's silver until 1880.

In the early 1870s, William Stephens and Alvinus Woods began a gold placer mining operation in Lake County at California Gulch. This gulch had been the site of an early 1860s placer district that had produced more than five million dollars in gold, despite a persistent problem of sluices that were regularly clogged by heavy sands. The new placer operation encountered the same problem, but Stevens recognized the sands to be largely composed of silver-bearing cerussite (a lead carbonate mineral) that was being profitably mined across the Mosquito Range in the Alma District. The partners located the deposits from which the sands had eroded and staked claims. Within a few years, Leadville became Colorado's largest silver-producing district.

Leadville, the quintessential western boomtown, is the stuff of legends. The population grew from 500 in 1877 to 30,000 in 1879, with 667 operating businesses—including 58 saloons. The Leadville District is a world-class mineral deposit with total silver production exceeding 260 million troy ounces. The list of famous Coloradoans with ties to Leadville, from Horace Tabor to Soapy Smith, is very long and includes five U.S. senators and five Colorado governors.

Several prospectors studying the preliminary geologic maps of the Hayden Survey noticed that the host rock to the Leadville ores cropped out on the western side of the Sawatch Range in the Roaring Fork River valley. During a prospecting trip to the valley in the summer of 1879, they located several claims near present-day Aspen, and the state's second-most-important silver district (more than 100 million troy ounces) was born. Eventually, prospectors discovered that a belt of silver deposits extended from Aspen Mountain, under the town, and up Smuggler Mountain. Today's Aspen Mountain ski lift system closely follows the route of aerial tramways used to transport ore from mine to mill.

The first silver deposits in Colorado were found in the vicinity of Georgetown, Clear Creek County. Problems arose in smelting the ores, and mining was not profitable. It was not until 1867, when Hill's Boston and Colorado Smelter was blown-in, that recovering silver became economically possible.

Soon after being discovered in Georgetown, silver was found just up the road at what would become known as Silver Plume in Clear Creek County. Mining developed slowly at first, but by 1868, many of the initial problems had been solved, and silver production increased sevenfold.

Several of Silver Plume's most prominent mines were on Republican Mountain, above the town. To the lower left just above town is the Diamond Tunnel, an access tunnel to the deeper portions of several veins. To the far left at the first switchback is the Pelican tunnel. To the far right at the second switchback are the Dunkirk and the Pay Rock.

January, the burro, poses underground at the Mendota mine in Silver Plume. She is hauling a mine car and carries two lanterns strapped onto her back. January has a carbide lamp around her neck, and another carbide lamp is on the ore car. An electric lightbulb, wiring, and insulators on the mine back (ceiling) help illuminate the way.

The east side of the Argentine District was several miles up the road from Silver Plume, while the west side was across the Continental Divide. This funicular tramway was a great labor-saving device at the Stevens mine. Before it was built, miners sewed up their ore in rawhide sacks and rolled them down the hill. Shrewd businessmen, the Stevens mine owners persuaded George Armstrong Custer to invest in the mine and used his name to help publicize their stock sales.

Caribou, at nearly 10,000 feet elevation in western Boulder County, was the second-most-important silver mining district during Colorado's territorial days. Logs were propped up against many of the buildings to keep them from blowing over. Caribou was known as the place where the winds were born.

The Caribou mine, on the crest of the hill, was acquired by Abel Breed in 1871. In 1873, he sold the mine for $3,000,000 to a Dutch syndicate, but while ships sailed across the Atlantic delivering the documents of the sale, Breed continued to mine the Caribou. Before the sale was finalized, he had mined out all of the developed ore bodies. The syndicate had spent all of its capital on the purchase of the mine, so it did not have money to find, develop, and mine new ore bodies. Thus, the syndicate was forced into bankruptcy within a few years. Jerome Chaffee and David Moffat, well-known Colorado mining men, bought the mine at a sheriff's sale for a song. Their capital went into development and mining, and the Caribou paid handsomely for a decade.

Five men are at work in the sorting room at the Caribou Mine. The journey from Caribou down to the company's mill in Nederland was long, and the freight costs were high, so ore was hand sorted to make sure its grade was high enough to yield a profit.

Under Abel Breed, the Caribou Mining Company constructed a mill in Nederland. The mill used the Washoe process, a less-than-perfect treatment method that had been developed at the Comstock Lode. The Nederland ores were rich (especially after hand sorting) and the mill was an economic success, despite recovery of not more than 80 percent of the silver.

This is probably the most famous Colorado mining photograph. Breed mined out so much ore, so fast, that his mill wasn't able to process it all. So, he sent ore to Nathaniel Hill's Black Hawk Smelter, where, as seen in this view, 30 silver ingots are stacked. Hill is standing in the doorway, and Richard Pearce, Hill's chief metallurgist, stands to the left with his foot on the boards. These ingots were used to pave a silver sidewalk for President Grant to walk on, from his stagecoach to the front door of the Teller House, during his 1873 visit to Central City.

In 1874, William Stevens and Alvinus Wood set up a hydraulic placer operation in California Gulch to recover gold that had been missed in the 1860 to 1862 mining efforts. They soon discovered that their sluices were clogging up with the same dark sands that had plagued earlier miners. Stevens recognized the sands as silver-bearing cerussite, the same mineral that was being mined on the other side of the Mosquito Range in the Alma District.

Stevens and Wood began a prospecting endeavor to find the source of the cerussite. By 1876, they had located and staked claims on outcrops along the western slope of Iron Hill. More prospectors began to arrive, building a new town and creating a new mining district, both named Leadville for the silver-bearing lead-carbonate mineral cerussite. This photograph was taken several years after Stevens and Wood's original claims were consolidated into the Iron Silver mine.

The Silver Cord, pictured here in 1881, and the neighboring Iron Silver were involved in a lawsuit over claim boundaries. The Silver Cord won, and judging from the elaborate super-structure that the company built, it held very rich ore.

In 1877, more prospectors arrived and discovered mineral deposits on Carbonate Hill, the next hill over from Iron Hill. By the time that William Henry Jackson took this photograph, the entire hill was covered with operating mines.

Three Irish emigrant brothers, John, Charles, and Patrick Gallagher, located one of the early Iron Hill mines. When the Gallaghers sold their mine for $225,000, the ensuing publicity brought a wave of Irish immigrants to Leadville, establishing a very large presence that remains today. The Maid of Erin mine, one of many Leadville mines with an Irish connection, is pictured in this photograph.

The climax of the Leadville silver discoveries came in 1878 on Fryer Hill. Horace Tabor, owner of a local general store, grubstaked two prospectors who located the Little Pittsburg mine. For a couple of years, the mine's production was phenomenal, making Tabor a millionaire, and propelling him on to other mining ventures. In this view of Fryer Hill, the Little Pittsburg is slightly left of center—a trestle runs from the shaft house out to a mine dump.

The Robert E. Lee mine may have been the richest mine on Fryer Hill. More than $250,000 was produced during a 39-day period in 1879, and another of its ore bodies produced almost $120,000 during a 17-hour mining effort.

By 1879, the Leadville District's population had reached approximately 30,000. Discoveries were still being made, and the future looked bright for silver. This view of Harrison Avenue shows construction in the right foreground just across the street from the Little Church on the Corner Saloon. Pap Wyman's log cabin is still standing in the middle of the street down the block. Wyman had built his cabin before the town was platted, and it took quite a legal effort to get the cabin moved.

By the 1890s, Leadville had matured. In this photograph, the viewer is looking west down 8th Street toward Mount Massive. The Coronado mine dumps are just to the left (south) of the street. In the left background, several smelter chimneys are visible. Irish life in Leadville was centered on the Catholic Church of the Annunciation—its location is marked by the tall steeple to the mid-left in the photo.

Hyman's Saloon on Harrison Avenue is just a few doors down from the Tabor Opera House. After his flight from Tombstone, Doc Holiday settled in Leadville in 1884. Holiday dealt Faro at the Monarch, Board of Trade, and Hyman's saloons. While at Hyman's, he was involved in a shooting and stood trial for assault with intent to kill. He was acquitted, left Leadville, and traveled to Denver, where he stayed for a couple of years. He then moved on to Glenwood Springs, where he died of consumption in November 1887.

This is the gaming room at the Pioneer Bar, one of Leadville's longest-lived entertainment establishments. Several gamblers are playing roulette at the front left. On the right, a Faro game is under way—in addition to the board layout, both the bank and the case keeper are plainly visible. The dealer is seated against the wall, and the banker is seated on a high stool to the dealer's right. The player directly across from the dealer is "keeping case," or keeping track of the cards as they are played. Two other players are at the far end of the table.

A large crowd has gathered at the intersection of Fifth Street and Harrison Avenue to watch the double-jacking event at the July Fourth drilling contests. At the 1888 Miners' Union Drilling Contest, Nicholas Myers and Thomas Rinker won $100 in gold for drilling a 30 and 3/8-inch-deep hole in 15 minutes.

Silver-lead ingots at a Leadville smelter are readied for shipment to a refiner, where they will be separated into the pure metals. The price of silver had been steadily sliding downhill for more than 20 years. Following the silver panic of 1893, the government ended bimetallism and halted the purchase of silver for coinage. Gold and base metals—lead, zinc, and copper—became much more important products of Leadville mining.

The Durant mine was located in what became the Aspen District in 1879. The neighboring Aspen mine was located in 1881. As it turned out, the Aspen was mining the Durant's ore. The resulting lawsuit went on for years before the two companies agreed upon a compromise consolidation. The Durant, shown in this photograph from the late 1890s, was used as a tunnel site to reach the deeper ore bodies in the consolidated properties.

A young man and woman pose during an underground visit to the Durant mine in Aspen. Old superstitions among miners prohibited women from going underground. So, had Aspen already become a bastion of social justice in the 1890s?

The mines of Smuggler Mountain, especially the neighboring Mollie Gibson and Smuggler shown in this photograph, were extraordinarily rich in silver. The Smuggler, which was the Aspen District's leading producer, was purchased from the discoverers by David Hyman, Charles Hallam, and Abel Breed, who was also an owner of the Caribou mine.

The Mollie Gibson, the last of the great Aspen mines to be discovered, worked incredibly rich deposits that were quite similar to those mined in the neighboring Smuggler. A railroad siding made shipping ore easy.

In 1893 to 1894, both the Mollie Gibson and Smuggler mines became famous for mining some very large silver "nuggets." Until the next discovery, the previous one was considered "Colorado's largest silver 'nugget'." There is a great deal of misinformation about these "nuggets," which aren't nuggets at all—nuggets being chunks of metal exposed to erosion and stream action, becoming somewhat round by having their sharp edges and corners smoothed off. The silver masses in view are pieces of the original, which approached or exceeded a ton in weight. During mining, chunks broke apart, and then the miners had to cut and break them up more in order to hoist them to the surface. The final result was that several chunks ended up on the surface, with each chunk labeled with a description of where the pieces fit together.

Aspen exhibited Queen Silver at the 1893 Columbian Exposition. The ten-foot-tall statue was made from silver-plated pressed copper and other materials the artist thought appropriate. She held a silver dollar as a scepter in her right hand and a scroll with the words "Free Coinage" in her left. She sat in a chariot shaped like a ship's prow decorated with a stag's head. The queen rested on an ornate platform with a canopy over head, and the entire work was eighteen feet tall. After the close of the World's Fair, Queen Silver was returned to Colorado, where she was placed on exhibit next to a similar statue of King Coal at the Mineral Palace in Pueblo. Years later, when the Mineral Palace went bankrupt, both statues were scrapped and sent to the smelter.

John F. Campion reopened the Little Jonny mine and found that it had potential for significant gold ores. In 1891, he incorporated the Ibex Mining Company, so named because of a personal superstition that his most outstanding mines were named for members of the deer and antelope family. The superstition must have been true, because the Ibex became Leadville's leading gold mine with a total production in excess of $25,000,000.

These gold ingots (some from the Ibex mine, all from mines operated by John F. Campion) are proof positive of the renewed emphasis on gold mining after the Silver Panic. Those mines with gold-bearing ore were able to survive the crash of the silver market. Some mines containing plentiful high-grade silver ore were also able to survive, but mines that contained only low-to-medium-grade silver ore were forced to close.

UNDERGROUND HARDROCK MINING

(1859–1920)

Not since the earliest days of mining have men used a pick to simply whale away at a wall of rock. Two revolutionary advances in mining technology in the seventeenth century gave miners an edge—German miners began using an explosion of black gunpowder to break rock into manageable chunks, and Hungarian miners began drilling holes into rock to hold the explosive charges. Mining soon became a six-step process of drilling holes in rock, filling the holes with explosives, blasting down as much rock as possible, loading and hauling the rock to the surface, separating ore from waste rock, and then extracting the metals from ore. This six-step process generally explains what miners do today, but there has been a continuing stream of technological advances throughout the years that has greatly improved the method and efficiency of each step.

During Colorado's early mining days, black powder, a low-power explosive, was the only blasting agent available. Dynamite, a much more powerful explosive, was invented in 1866 and was in common use by the 1870s. During those black powder days of mining, holes were drilled in rock by hand. Miners struck a chisel-pointed drill steel with a sledgehammer and eventually drilled a dozen or so holes, 10 to 24 inches deep, into solid rock. Usually the process involved two miners who alternated between holding the drill steel steady and swinging an eight- to ten-pound hammer. In cramped spaces, a single miner held the steel with one hand and struck with a four-pound hammer.

By the mid-1870s, pneumatic machine drills had been developed, resembling the noisy jackhammers used today by road construction crews to break up concrete. With a machine drill and dynamite, a two-man team in 1875 was able to drill and blast down four to ten times as much rock in a shift as a two-man team in the 1860s.

The years from 1860 to 1920 mark the flowering of the Industrial Revolution. An 1865 Central City gold miner transported to a World War I–era tungsten mine would still recognize the basic six-step mining process. He would marvel, however, at the changes in the support technologies involved in mine lighting; generation of power; underground haulage and hoisting of men, machinery, and ore; timbering and support of underground workings; water pumping and drainage; ventilation; ore treatment; transportation to the smelter; and the extraction of metals.

A miner uses a pick to pry down loose rock and clean up the newly blasted opening, or portal, to his mine.

All mines start small—machinery comes with time and economic success. In this photo a miner hoists rock with a windlass. The shaft might have been only 10 or 20 feet deep, and access for work would have been by ladder.

In this early view of the Buena Vista mine's surface plant in the Cripple Creek District, things appear to be progressing nicely. The shaft is undoubtedly deeper and now requires a one-horsepower windlass to hoist rock. Miners have built a headframe to support a sheave wheel assembly (working as a simple pulley), as well as an ore-loading house. A log shaft house is under construction. The Buena Vista went on to become a key mine in the Isabella Gold Mining Company's consolidated holdings with a total production of more than $15,000,000 in gold.

This relatively simple steam-powered hoist uses a bucket to hoist men, equipment, ore, and waste rock. A miner would ride the bucket by standing on the rim and holding onto the cable. As many as three men, arranged to balance the load, could ride the bucket to work. Even a ride at slow speeds was somewhat of an athletic endeavor, because the bucket would twist and turn during the trip up or down the shaft.

In a photo of the hoist room at the Cook shaft, Central City District, the hoist engineer operates a twin-cylinder, double-drum hoist for a twin compartment shaft. Just out of the field of view a sign warns, "Do not talk to the hoist engineer"—for good reason. The twin cylinder hoist can move men, rocks, or equipment at more than 2,000 to 3,000 feet per minute (approximately 35 m.p.h.). An arm rotates around each of the round dials facing the engineer. The position of the arm indicates the depth of the cage in the shaft.

In the Central City District, six miners stand in an open steel cage at the shaft collar, waiting for the hoist engineer to lower them down the Cook shaft.

A miner is turning a windlass to lift a bucket of ore from a stope through a raise or winze. Raises and winzes are internal shafts that have been driven up, or sunken, from a haulage level to a stope, where ore is actually mined. After the ore is broken up by explosives, it is shoveled into a bucket and lowered or lifted to a haulageway, where it is loaded into an ore car and taken to the surface.

The area from which ore is actually extracted is called a stope. In this view of a stope, one miner is operating a pneumatic drill, known as a stoper. His supervisor (holding a light) looks on, and a third miner is climbing up a ladder from a raise into the stope.

Two miners are hand drilling, or single-jacking, on platforms in a large stope in the Cripple Creek District. In single-jacking, a miner would stand on a narrow platform and pound a hole into solid rock by striking the drill steel with his hammer. Several candles provide light for these miners.

It was common for miners to work in cramped quarters. In this view, a miner is single-jacking by candlelight.

At the Saratoga mine in the Central City District, a miner swings his single-jack hammer as he drills into the back of a narrow vein. Working by candlelight, he stands on a plank supported by what is known as stull timbering—a single beam held in place by wooden wedges and notches chiseled into the rib. The ceiling of a mine is called the back, and the mine wall is called the rib.

Working by candlelight in a large stope, three miners are single-jacking, and one team of miners is double-jacking. The term single-jacking comes from the Cornish, whose miners were known as Cousin Jacks. Single-jacking was done by one miner, double-jacking by a team of two miners—one holding and rotating the drill steel, and the other striking the drill steel with a double-jack hammer (essentially a sledgehammer).

Long after machine drills had largely replaced hand-drilling methods, miners still kept the old traditions alive with drilling contests for substantial amounts of prize money. In this photo, a team of miners is double-jacking in a Fourth of July drilling contest at Creede: one miner swings a sledgehammer to strike a hand steel, while his partner steadies and rotates the steel one-quarter turn after each blow.

A miner is drilling up into the back of a vein with a pneumatic drill. The drill that he is using is known as a drifter; it would normally be used for drilling horizontally oriented holes.

The miner in this photo is drilling the vein with a stoper by candlelight. He stands on a drill steel wedged between the ribs, and he has anchored the leg of his stoper in a hitch chiseled into the rib. Machine drills came into use in the mid-1870s and greatly increased the amount of rock that could be drilled and blasted in a working shift.

A heavy steel column supports a drifter. One miner turns a handle that advances the drill steel into the drill hole. A hose carries compressed air to the drill, which is released by a valve to propel a piston in a cylinder. Essentially like a modern jackhammer, these noisy drills hammered a hole into the rock in a remarkably short time. The photo, taken at the Gold Coin mine in Cripple Creek, is obviously posed—the miners are "drilling" into a rock pillar purposely left in place to support the back from collapsing.

Two miners are loading drill holes with dynamite. The wooden pole leaning against the rib is used to push the dynamite to the back of the drill hole. Different fuse lengths will allow a timed sequence of blasts that will break the rock more efficiently than a single big blast.

After a round of dynamite was loaded and shot, the broken rock was shoveled into an ore car. Miners called the shoveling and loading process "mucking." After the broken rock was mucked out, the mine railroad tracks were extended, and the process of drilling, loading, and shooting was repeated.

Some mining engineers constructed a series of loading chutes in the stopes so that rock could be loaded more easily into ore cars. This photo was taken at the Saratoga mine, Central City District.

In this posed photo, a blacksmith and his helper sharpen drill steel. The bit end of the steel is heated red hot, a star-shaped die is placed and held over the bit by a helper, and the blacksmith hammers the die to pound the dull bit back into shape. Pneumatic drilling machines dulled dozens of bit ends on drill steels in a ten-hour shift. Large mines employed about half as many blacksmiths as miners.

Some rock is reasonably self-supporting, while some is not. Miners used various forms of timbering to brace non-self-supporting rock and prevent cave-ins. The square-set timbering shown in this photo is holding up bad ground, allowing miners to work in relative safety.

Rock bursts were a problem in some mines—pressure built up from the weight of the over-burden caused the rib to collapse in a shower of rock shrapnel. In this mine, engineers have installed square sets and then timbered the rib to eliminate danger from this problem.

In the 1890s, a horse pulls loaded mine cars at the Bobtail mine in the Central City Mining District. Stull timbering with working platform is visible behind the miner standing on the ladder.

Large mines could have miles of underground workings with an extensive rail tramway that included branching lines, switches, and sidings.

By the beginning of the twentieth century, electricity was in use at many mines. In this posed photograph, an electric trolley locomotive pulls a train of ore cars from the portal of the Bobtail mine to the Fifty Gold Mines mill located in Black Hawk, Central City Mining District. The stone building is still standing today.

Miners sorted various types and grades of ore into this row of bins at the Consolidated Gem Mine in Gilson Gulch, near Idaho Springs. One bin contained low-grade lead-silver ore that would be sent to a mill for concentration, another contained high-grade ore that would be shipped directly to a smelter, and yet another contained zinc-rich ore that would have to be specially processed. A teamster would stop his ore wagon at the appropriate bin and load it up.

The teamsters in their ore wagons seem to have a traffic jam on their hands in Idaho Springs. Until the arrival of the internal combustion engine, scenes like this were not uncommon in most mining towns.

It took some ingenuity and a lot of horses, mules, donkeys, and oxen to get equipment up the narrow mountain roads of frontier Colorado. A teamster at Dumont in Clear Creek County has two teams pushing and four teams pulling a boiler to a mine just up the hill.

In this photo, an electric trolley pulls a train of ore cars out of the Newhouse (Argo) Tunnel in Idaho Springs. Samuel Newhouse formed a company to drive a deep adit, or mine tunnel, more than four miles from Idaho Springs to Central City to drain the mines of the area, provide cheap haulage for ores, and bring those ores to his mill for treatment. The Newhouse venture was a financial failure, because mine owners figured out that their mines would be drained whether they paid the subscription charges or not.

Three succeeding companies eventually drove the Argo Tunnel, previously the Newhouse Tunnel, to a length of 21,968 feet. However, each failed to complete the overall project. The mill at the mouth of the tunnel charged exorbitant rates and could not perform the custom work necessary to handle some complex ores. Nevertheless, other deep drainage-haulage adits in Colorado, such as Leadville's Yak Tunnel, were successful, largely because the mining companies that owned the mines drove and operated the tunnel.

Most mining companies based almost all of their important decisions on information gained from assays. As shown in this photograph, an assayer poured a molten ore sample from a crucible into a conical mold. After the melt cooled, the heavy metals were solidified at the tip of the cone below the lighter nonmetals (e.g., quartz and feldspar). The metal-bearing tip was then broken off and re-melted in a cupel, which absorbed lead, zinc, and copper, leaving the gold or silver as a bead. The exact weight of the bead was then used to calculate the value of the ore in troy ounces per ton of ore.

Several assay balances for weighing the gold or silver bead from the cupellation process are prominent in this view of the assay lab at the Fifty Gold Mines Corporation mill in Black Hawk. Various chemicals and racks of laboratory glassware are also present in the lab.

The Perigo Mining Company's mill in Gilpin County was a medium-sized mill plant typical of those in the beginning of the twentieth century. It was built on a hillside so that gravity could be utilized to help move the ore through a series of concentrating processes: 100 tons of low-grade ore were crushed, and the metallic minerals were separated from the enclosing rock, yielding 17 tons of concentrate. The mining company was able to sell the concentrate to a smelter for a considerably higher price than the original 100 tons of low-grade ore.

The upright structure running horizontally across the center of this photograph holds eighty 800- to 1,000-pound steel stamps used to crush ore. Each stamp was power lifted and then allowed to fall on the ore, 93 times per minute. The actual stamps aren't visible because they are enclosed in a screen-covered box containing water and mercury. During crushing, the mercury mixed with gold and silver in the ore to form amalgam, which splashed (with the water) through the screen onto the slanted tables in front of the stamp batteries. The tabletops are sheets of copper coated with more mercury to trap the amalgam. During periodic cleanups, the amalgam was scraped up and heated in a retort to vaporize the mercury, which was piped off and condensed, and then collected for reuse. The gold "sponge" left in the retort was sent to be refined. The major refiner in Colorado was the U.S. Mint in Denver.

Stamp mills with amalgamating tables didn't recover all the gold and silver from complex ores, so metallurgists designed and installed many machines in a circuit that systematically processed the ore and concentrated the metals. In this photo, the millman leans against a set of jigs, while the man on the right stands by a shaking table. Both machines utilize gravity to separate the heavier metallic minerals from the lighter gangue minerals, such as quartz, feldspar, and calcite.

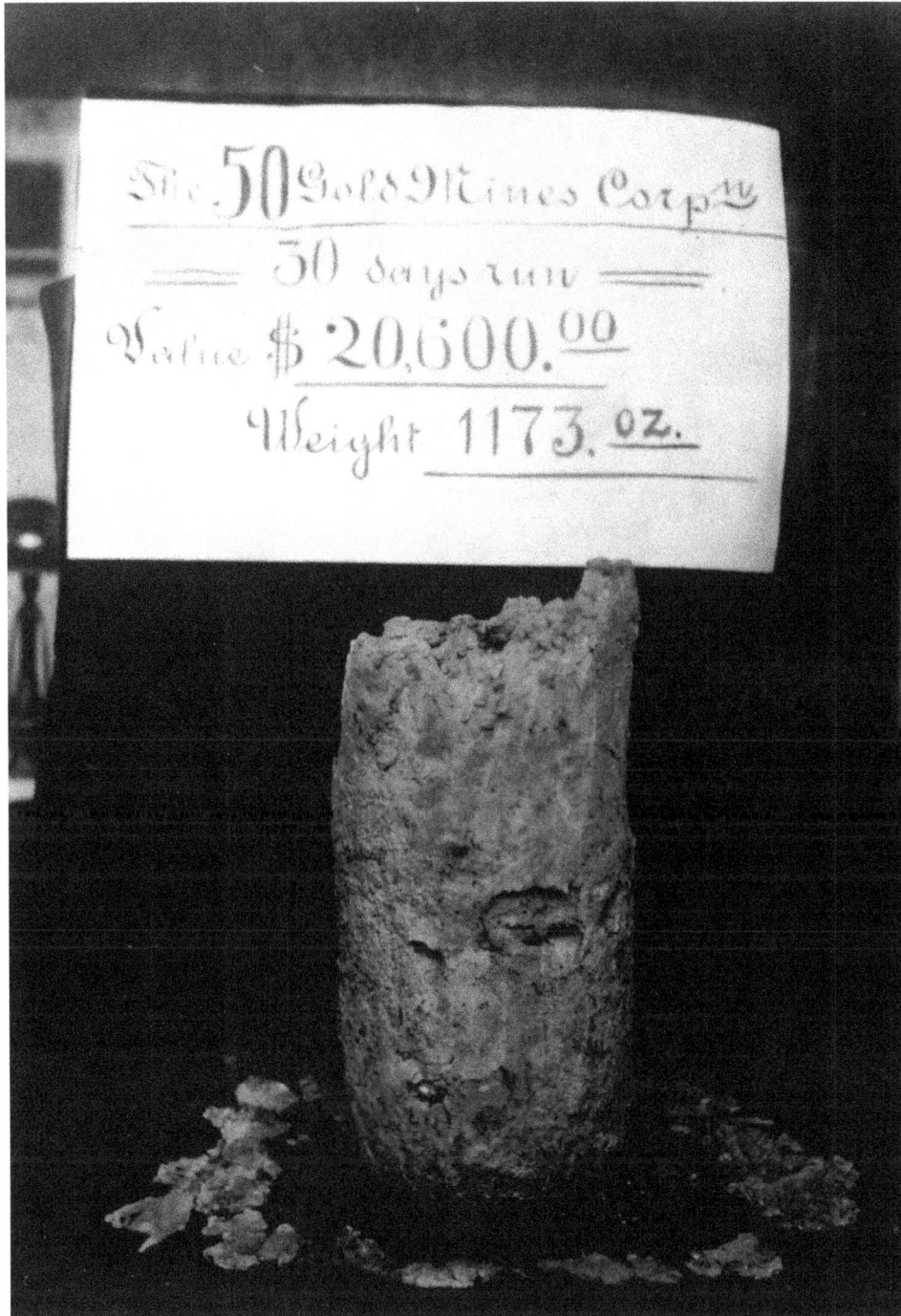

The 50 Gold Mines Corp^y

50 days run

Value $ 20,600.^{00}

Weight 1173. oz.

In this photograph, the Fifty Gold Mines Corporation proudly displays an 1173-troy-ounce gold sponge valued in its day at $20,600, recovered from its retort. This sponge is the end product of a month's mining and milling. The Fifty Gold Mines mill could process 400 tons of ore per day, so the ore averaged close to one-tenth of an ounce of gold per ton of rock.

A mining company's final step in metals recovery was to send the concentrated ore from its mill to the smelter. The Omaha and Grant Smelter north of Denver is shown in this photo.

Mining was dangerous work. In August 1895, 14 Central City miners drowned when, in just minutes, water filling the long-flooded-out lower workings of the Fiske mine broke through and flooded portions of the neighboring Sleepy Hollow and Americus mines. This photo shows horse-drawn hearses and carriages queuing up for funeral services at St. Mary's of the Assumption Catholic Church.

THE SAN JUANS

(1871–1920)

The San Juan Mountain region in the southwestern quarter of the state is one of the world's great volcanic fields. In the nineteenth century, these lofty mountains were considered remote. Today, words like spectacular, incredible, and the much-hackneyed "awesome" are regularly applied to this land of jagged peaks. Originally home to the Ute Indians, the area was officially opened to prospectors with the Brunot Treaty, signed just a few years after they had discovered mineral deposits there. The three towns of what is known as the "San Juan Triangle," Silverton, Ouray, and Telluride, were all founded in 1874–1875. Summitville and Lake City were settled earlier, Bonanza and Rico later. Creede, a story unto itself, was founded much later in 1890.

Each of these towns owes its existence to mining. Largely because of the rugged topography, mining developed slowly, and the region's growth was spurred by, and became largely dependent upon, the narrow-gauge railroad. Telluride and Rico were served by the Rio Grande Southern. The Denver and Rio Grande provided service to both Ouray and Silverton, but because of incredibly rugged terrain, there was no direct rail connection between the two. The D&RG also served Lake City and extended its tracks from Wagon Wheel Gap to Creede in the 1890s. Three short lines connected Silverton with several nearby mining areas. Tourists to the area today can still take the train from Durango to Silverton—one of the nation's most spectacular train rides.

Silverton got its name because the initial discoveries were not gold deposits, but as the prospectors said, "We've got silver by the ton." Silverton's name notwithstanding, six rich gold mines accounted for a major share of the mineral production in the San Juan Triangle: the Camp Bird at Sneffels near Ouray; the Sunnyside and Gold King near Silverton; and the Tomboy, Smuggler-Union, and Liberty Bell near Telluride. One of the more unusual deposits in the San Juans was Colorado Fuel and Iron Company's fluorspar mine at Wagon Wheel Gap. Fluorspar was shipped to Pueblo for use as a flux in the CFI steel mill.

The Ute and Ulay veins, discovered in 1871 by Henry Henson, J. K. Mullen, Albert Meade, and Charles Godwin, were located just a few miles up Henson Creek from a valley that would become the site of Lake City. Although the veins were the first significant discoveries in the area, the land belonged to the Ute Indians. As more prospectors drifted into the area, the U.S. government purchased a 60-by-75-mile strip of land from the Utes as part of the Brunot Treaty, drawn up to avoid open war. In 1876, the Ute and Ulay were purchased by the Crooke brothers, Jonathan J. and Lewis, and consolidated as the Ute-Ulay. Water was supplied to their mill through a flume, seen running across the lower field of view in the photo.

Enos Hotchkiss staked the Hotchkiss lode in 1874, initiating a mini–gold rush that brought some 400 prospectors to the Lake City District area. The Hotchkiss lode was sold several times and was eventually renamed the Golden Fleece mine. In this photo, more than 50 miners and two cooks pose for the camera.

In the 1870s, J. J. and Lewis Crooke operated the most successful smelter in the San Juans. The Crookes competed successfully with other smelters until the 1882 arrival of the Denver and Rio Grande Railroad in Silverton, which made John Porter's smelter at Durango virtually invincible in the smelter war.

In 1874, a stage road was built over Cinnamon Pass, winding through Sherman, Burrows Park, and Whitecross on the way. The Black Wonder mill stands behind the Sherman general store in this photograph.

At the beginning of the twentieth century, Silverton was at its prime. This thriving mining town had narrow-gauge rail connections with all of the surrounding major mining and milling centers, including Gladstone, Arrastra Gulch, Howardsville, Middleton, Eureka, Animas Forks, Chattanooga, Red Mountain, and Ironton. As evidenced in this photograph, Silverton also had a beautiful backdrop, with Sultan Mountain (highest peak on the right) and Grand Turk (highest peak on the left) standing as sentinels above the valley floor.

Railroads were the key to mining in the rugged San Juan Mountains—wagons and burro pack trains were simply an insufficient means of commercial freight transportation for rocks and heavy equipment. In this 1890 view, the Denver and Rio Grande narrow-gauge locomotive number 61 pauses for water—and a photo—at Needleton, between Durango and Silverton.

Two engines of the Denver and Rio Grande pull this train along cliffs high above the river in Animas Canyon between Durango and Silverton.

Areas where streams have cut channels down steep mountainsides fill with snow during winter and become avalanche sites. At some places in the San Juans, these chutes channel avalanches with such regularity that they are said to "run," in almost the same sense that streams run through the channels in the summer months. Some chutes have even been given names—this 60-foot-deep pile of snow in Animas Canyon is the Saguache Snowslide.

A regional smelter at Durango was made possible by the railroads. This view is titled "American Smelter," which dates the photograph to sometime after 1899. However, it is the same plant—with the exception of improvements—as the old San Juan and New York Mining and Smelting Company, first managed by John Porter in 1880. Throughout the years, it was reorganized and went through two mergers before being acquired by the American Smelting and Refining Company.

The end-of-the-line customer for the Silverton Northern Railroad was the Golden Prince mill at Animas Forks.

One of San Juan County's most famous mines was the Sunnyside, located at approximately 12,500 feet above the community of Eureka. This view shows the large surface plant with Lake Emma in the foreground.

The aerial tramway between the Sunnyside mine and mill was three miles long and was used as an angle station to bypass a bend in the route. A one-way trip from mine to mill took about 45 minutes. This view was taken from inside the tramway station.

The 500-tons-per-day milling facility of Eureka's Sunnyside Mining Company was just a few miles above Silverton on the Animas River. It was built on a slope in stair-step fashion to allow the ore to move easily from level to level and process to process.

Fire was a constant threat for the wooden mining camp structures. The April 24, 1919, fire at the Sunnyside mine destroyed most of the physical plant at a loss of $400,000. The fire followed closely on the heels of the 1918 flu epidemic, which killed 14 percent of Eureka Mining District's population.

The *Casey Jones,* a special engine and car combination, was built by the Sunnyside Mining Company for use by mine personnel traveling between Eureka and Silverton on the Silverton Northern Railroad. In service during the Judge Terry years, it was powered by a Cadillac automobile engine. This strange little vehicle sits on display in Silverton.

Located on King Solomon Mountain, this portal of the North Star mine, owned in part by the Crooke brothers, sits nearly 13,000 feet above sea level. It was not uncommon for several mines to have the same name, especially if the name was derived from a well-known landmark, person, or event. Not nearly as well known as California's famous North Star gold mine, the King Solomon North Star did outshine Silverton's other North Star mine on nearby Sultan Mountain. Much of the ore from the North Star was shipped by burro train over the mountains to be smelted at the Crooke Smelter at Lake City.

Dozens of San Juan mines were perched in seemingly impossible places among the mountain crags. San Juan County's Old Hundred mine is one of the most spectacular.

Mining is hard labor, and the hours were long in the late nineteenth century. Add into the equation life in an isolated boardinghouse at high altitude, and a crew of San Juan miners could become a temperamental workforce. Wise mine managers did their best to provide a few comforts of home, such as good food—and plenty of it. The two cooks are pictured here with the miners and the rugged cliffs crowding the boardinghouse. This unidentified photo is probably the boardinghouse at the Old Hundred mine.

A pack train of burros hauls mine timbers to the "White House," the Highland Mary mine office and boardinghouse for the mine in upper Cunningham Gulch. Founded in 1876 by a wealthy, eccentric New York spiritualist named Edward Innis, the mine was developed by orders from what he deemed "the spirit." Evidently, the spirit didn't understand the economic necessity of mining and shipping ore, and Innis used up his capital without mining the ores he found. Broke, he returned to New York and died insane.

In the years between 1910 and 1920, Telluride was at its peak. One of the roads to the mines is visible above town, as is the Pandora milling center toward the head of the valley. Ingram Falls can be seen pouring its water into the valley from Ingram Basin, above.

Between 1875 and 1880, the Humboldt, Mendota, Sheridan, Smuggler, Union, Cleveland, and Cimarron claims were staked on the Smuggler vein in Marshall Basin. In this photograph of Marshall Basin taken from the Mendota mine around 1889, the surface plants of the Sheridan and Smuggler are visible. Below and running parallel to the mountain ridge on the right is the Smuggler vein. Across the canyon, another vein can be seen running up the mountain.

A syndicate controlled by Ernest Waters acquired the Sheridan mine in 1883 and turned the operation into the first real economic success in the Telluride District.

Ernest Waters constructed an 8,400-foot-long, gravity-powered, inclined, rail tramway from mine to mill on the valley floor. The double-tracked tramway was equipped with an endless cable, which allowed the downhill-bound loaded cars to unload at the bottom and be pulled back up to the mine. The route, which crossed nine bridges and passed through four tunnels, was considered one of Colorado's mining wonders.

In this photograph of Pandora, the abandoned Sheridan Incline can be seen running up the hill to the left, and the first of the Smuggler Union mills stands on the lower hillside to the center right.

In 1891, all of the Smuggler vein properties were consolidated into the Smuggler Union Consolidated Mining and Milling Company, with John Porter, former manager of the Durango smelter, serving as company president. In 1899, the mine was sold to the Livermore-Agassiz-Shaw syndicate, a group of mining men who controlled the great Calumet and Hecla copper mine in Michigan. In this view, much of the Smuggler Union surface plant can be seen spread out across the mountainside.

Perched picturesquely above Telluride is the surface plant at the Bullion Tunnel level of the Smuggler Union mine. Directly behind the Bullion Tunnel plant is Savage Basin, dotted with several buildings of the Tomboy mine.

As shown in this photograph, the surface plant at the Bullion Tunnel level of the Smuggler Union mine clings to the mountainside. The building with the long row of dormer windows along the roof is the miners' boardinghouse. The long curving structure in the foreground is the snowshed-covered upper-tramway station, with a tram tower visible about 150 feet down the slope.

IS COLORADO IN AMERICA?

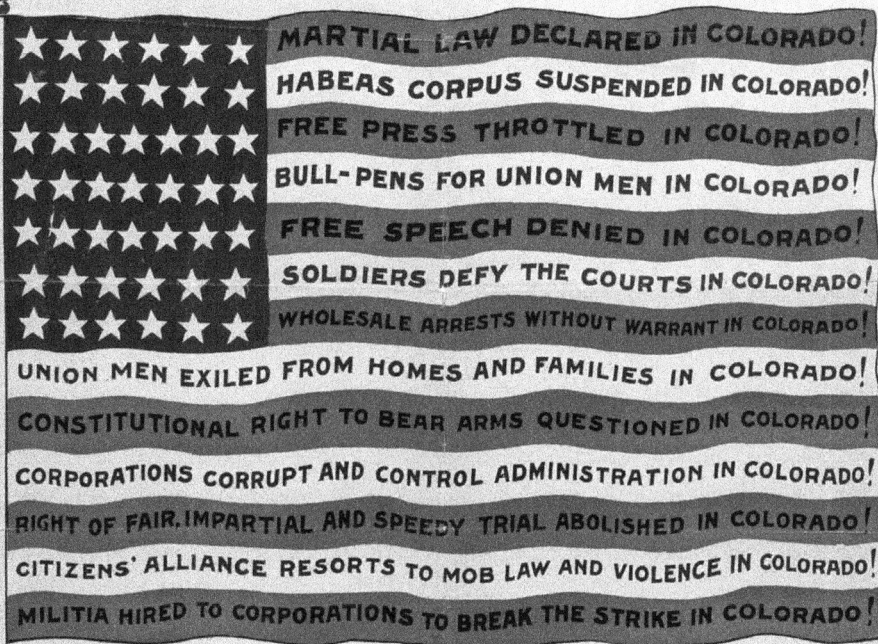

MARTIAL LAW DECLARED IN COLORADO!

HABEAS CORPUS SUSPENDED IN COLORADO!

FREE PRESS THROTTLED IN COLORADO!

BULL-PENS FOR UNION MEN IN COLORADO!

FREE SPEECH DENIED IN COLORADO!

SOLDIERS DEFY THE COURTS IN COLORADO!

WHOLESALE ARRESTS WITHOUT WARRANT IN COLORADO!

UNION MEN EXILED FROM HOMES AND FAMILIES IN COLORADO!

CONSTITUTIONAL RIGHT TO BEAR ARMS QUESTIONED IN COLORADO!

CORPORATIONS CORRUPT AND CONTROL ADMINISTRATION IN COLORADO!

RIGHT OF FAIR, IMPARTIAL AND SPEEDY TRIAL ABOLISHED IN COLORADO!

CITIZENS' ALLIANCE RESORTS TO MOB LAW AND VIOLENCE IN COLORADO!

MILITIA HIRED TO CORPORATIONS TO BREAK THE STRIKE IN COLORADO!

UNDER THE FOLDS OF THE AMERICAN FLAG IN COLORADO!

EVERY WORD inscribed upon the stripes of "Old Glory" is the truth. If this flag is desecrated, the Republican Governor of Colorado is responsible for the acts that profane the emblem of liberty.

THE PICTURE represents Henry Maki, a union miner of Telluride, who was arrested for vagrancy---had money in his pocket and was being supported by his union. He was shackled to a telephone pole because he refused to work in a filthy cess-pool under the bayonets of the state militia.

WE ARE GOING TO BREAK his chains and the chains that are binding the working class of Colorado.

OUR STRUGGLE is for an eight-hour day, to establish the right to organize for mutual benefit, and to prevent discrimination against union men.

IF YOU DESIRE to assist the striking Miners, Mill and Smeltermen of the Western Federation of Miners of Colorado in this battle for industrial and political freedom, send donations to Wm. D. Haywood, Sec'y-Treasurer, 625 Mining Exchange Building, Denver, Colorado.

Charles Moyer.
PRESIDENT

Wm D Haywood
SEC'Y-TREASURER

The miners' strikes at Telluride were especially bitter, with several gun battles, an assassination, dynamite bombings, mass deportation of scab laborers by strikers, the calling out of the Colorado National Guard, imposition of martial law, and mass deportation of strikers by the state militia. Edward Boyce, president of the Western Federation of Miners (WFM), declared that labor should receive every dollar of wealth that it produced. In 1904, the WFM circulated this poster showing Telluride miner Henry Maki handcuffed to a telephone pole. After being convicted of vagrancy, five men were ordered by the sheriff to a work project that consisted of shoveling a cesspool into an excavation. Maki refused, so he spent one hour and 25 minutes shackled to the telephone pole.

The San Juan District [Mine] Owners Association built Fort Peabody on Imogene Pass to guard against Western Federation of Miners sympathizers sneaking into the District.

Snowsheds cover the mine railway from the Stillwell Tunnel to the aerial tramway station of the Liberty Bell mine. The mine boardinghouse and several other structures are visible in the background.

The Tomboy mine and mill were located in Savage Basin in the Telluride District. The property, the most profitable of the Telluride District, was controlled by the Rothschild interests through the London Venture Corporation. The Japan mine can be seen in the background.

A bucket starts its way down the hill from the upper aerial tramway station of the Tomboy mine.

The aerial tramway cables of the Smuggler Union and Tomboy cross each other as shown in this photograph.

Aerial tramways were often the most practical conveyance for supplies to be sent uphill. A bundle of mine timbers has been secured to two tram buckets and is en route to the Smuggler Union mine.

A pack train makes its way down the switchbacks of the Smuggler Union Trail. Near the bottom of the photograph, a snowshed-covered mine railway from the Pennsylvania Tunnel portal crosses over the trail.

In this view of Pandora, the large building to the right is the "Old Red Mill" of the Smuggler Union, the large building toward the front is the Ball mill of the Smuggler Union, and the large structure to the left rear is the mill of the Pandora Gold Mining Company.

The town marshal leans against a deserted bar, and the roulette wheel is also deserted—everyone is playing Faro, by far the West's favorite game of chance in the nineteenth and early twentieth centuries. The Cosmopolitan, next door to the Bank of Telluride, was a first-class saloon and gambling house.

Dave Woods, freighter par excellence, strung coils of a 10,810-foot wire rope from mule to mule on both sides of the animals. With the load balanced, 52 mules made their way out of Telluride and up the trail to the Nellie mine. The rope would be used for the mine's aerial tramway.

Dave Woods's mule train is on the trail, bound for the Nellie mine with 10,810 feet of wire rope.

Traveling on the "Rainbow Route" of the Silverton Railroad, a freight train pauses at the National Belle mine in the Red Mountain District of Ouray County.

In the right foreground of this view, a narrow-gauge railroad siding serves the ore house of the Yankee Girl mine in the Red Mountain District. Immediately above the ore house is the shaft house (still standing today), and to the left are the engineering building, with smokestacks from the boilers, and the coal house. The complete shift crew—both miners and surface workers—appears to have turned out for the photograph.

Just north of the town of Ouray, the American Nettie mine clings to cliffs of the gold-bearing Dakota Sandstone.

This view shows the upper Camp Bird 3 Level adit in Imogene Basin, at an altitude of approximately 11,400 feet. The miners' boardinghouse is to the right, the upper tramway station and ore-loading facility is in the center, and a tramway tower is to the left. The mine portal is outside the field of view to the left, at the end of the mine railway trestle. Tom Walsh discovered gold in some old mine workings in 1896, and he quickly consolidated his holdings along the length of the vein. The Camp Bird became the greatest gold mine in the San Juans; during its peak years, 1900 through 1912, the mine gained a profit of $17,731,788.

Tom Walsh took special care of his Camp Bird miners. The dining room at the mine's boardinghouse was well-known for serving the best food of any boardinghouse.

Here at an altitude of 12,000 feet, a mule pack train crosses a snowslide zone in Imogene Basin. Winter brought a real and constant avalanche danger to the high country of the San Juans, but the mines needed supplies, and a pack train was sometimes the only way to get them there.

In 1906, an avalanche and fire destroyed the Camp Bird mill in Imogene Basin. Shown here on this snowy day is the aftermath of both disasters.

In 1902, Tom Walsh sold the Camp Bird mine to Camp Bird Limited. After fire and avalanche destroyed the mill in 1906, Camp Bird Limited completely rebuilt the facility with the latest ore-processing equipment.

Sometimes avalanche rescue parties only recovered bodies. This body is strapped onto a sled for the return trip home (ca. 1890s).

In the 1890s, a small community formed in the vicinity of the Revenue Tunnel and mill.

The Virginius mine was located in Governor Basin at 12,180 feet above sea level. A. E. Reynolds acquired the property and operated the mine through his Caroline Mining Company. It took Reynolds's crews four and a half years to drive the Revenue Tunnel to an intersection with the Virginius vein.

THE LAST TWO RUSHES

(1889–1920)

In 1890, the U.S. Bureau of the Census declared the frontier to be closed. Although both the Creede and Cripple Creek discoveries date from the last days of the frontier, both districts reached their apex in what the Census Bureau identified as the "modern era."

John McKenzie staked the Alpha claim in 1876 at a site that would become known as Sunnyside, approximately three miles west of present-day Creede. The mine usually paid expenses, but it was not rich enough to attract great attention. In 1889, Nicholas C. Creede prospected East Willow Creek and found an outcrop of the silver chloride mineral, chlorargyrite, causing him to exclaim, "Holy Moses, chloride of silver!" He named his claim the Holy Moses and worked it in secret until he had accumulated enough ore to ship. In 1891, prospectors located the Last Chance, Amethyst, and Commodore.

Mining man and banker David Moffat convinced the Denver and Rio Grande Railroad that an extension of track from Wagon Wheel Gap to the newly established town of Creede would be a paying proposition. Unlike Colorado's earlier gold and silver rushes, the Creede Rush was conducted by rail. The first train arrived in December 1891 with hundreds of passengers. For the next several months, between 150 and 300 arrived by train every day to this town, where well-lit saloons, gambling houses, and bordellos operated around the clock. Newspaper editor Cy Warman seems to have had this phenomenon in mind when he wrote in his poem "Creede" that "It's day all day in the day-time / And there is no night in Creede."

But night did come. The steady 20-year decline in the price of silver reached a critical point in 1893, when a panic hit the markets and marginal silver mines were forced to close. Suddenly gold was the metal of choice. Cowboy Bob Womack had discovered gold in 1889 at a place that would become known as Cripple Creek. Like Leadville, Cripple Creek is a world-class mineral deposit; however, it is an unusual deposit. Almost all the gold is contained in minerals consisting of gold and silver in combination with the element tellurium. Worldwide occurrences of this type of gold deposit can be counted on one hand.

Cripple Creek has produced more gold than all the rest of Colorado's mining districts put together, and it is still producing. Like Leadville, the Cripple Creek story is also the stuff of legends.

In 1876, John McKenzie discovered silver near the confluence of Miner and Rat creeks in what would become Mineral County. In 1889, Nicholas Creede (born William Harvey) discovered silver on East Willow Creek and staked the Holy Moses; and in 1890, David Moffat, a well-known mining man and banker, bought the Holy Moses. Moffat's name aroused interest in the discoveries, and the rush was on. A camp called Willow sprang up but was renamed Creede in October of that year. One of the first businesses to arrive during any frontier rush was a saloon, and the first one was usually just a tent like the one pictured here. The next one might add a long board balanced on two barrels and call it a bar.

David Moffat, along with others, convinced the Denver and Rio Grande Railroad to extend its tracks to Creede from its end-of-the-line station at Wagon Wheel Gap. The first train pulled into town on December 10, 1891. People arrived daily by the hundreds, ore was shipped out by the tons, and merchandise and mining machinery arrived by the tens of tons. According to several photos, some folks even rode into town on the train roof—a common practice today in India.

Everything was fast-paced during the Creede boom—buildings went up in record time, and the population grew rapidly. The streets were clogged with people, wagons, piles of lumber and logs, kegs of nails, draft animals by the dozen, mud, and manure. On February 1, 1892, John W. Flintham decided the town needed electric lights, and five days later the lights were turned on at a little after 11:00 P.M. Later that year, Cy Warman, editor of the *Chronicle,* wrote a two-stanza poem in which he proclaimed, "It's day all day in the daytime / And there is no night in Creede."

Before the railroad arrived, the mail was delivered to Wagon Wheel Gap and brought to Creede by wagon. There, as historian Nolie Mumey put it, the mail was dumped "on the floor in a corner of a rough shanty, about ten feet square, to be sorted. There [was] a counter loosely nailed together; behind this stood the postmaster, knee deep in mail."

Some photographs of boomtown Creede look like noon at Midway on the State Fair's opening day. Others, like this one, have more of a morning-after look to them. The Silver Panic of 1893 ended the Creede boom, but through good times and bad, mining was an important part of life at Creede for a hundred more years.

The modern-day visitor to Creede sees a town confined to the valley floor in front of the mountains, but the frontier town ran all the way up the canyon past the forks of East and West Willow creeks.

Without question, one of Colorado's most spectacular mining scenes today is the Commodore mine, its levels strung up the mountainside above West Willow Creek. This is the way the scene looked 100 years ago after A. E. Reynolds had acquired the property. Reynolds operated the mine through a series of adits, or tunnels, driven into the hillside. The Number 5 Tunnel ran through the length of the claim and connected with the 1,200-foot level of the Last Chance and Amethyst mines.

Miners pose with several trains of mule-drawn ore cars at the Commodore 3 Level Adit (Tunnel). Ore from the 3 Level was transported downhill by aerial tramway to the 4 Level Adit (Manhattan Tunnel), where it was combined with ore from this level then transported on the 4 Level aerial tramway to the railroad on the canyon floor.

The Commodore began operations in April 1891, but it did not really come into its own until several years later. The surface plant at the Commodore 3 Level Adit was built on a leveled rockfill supported by log cribbing. The 3 Level aerial tramway can be seen in this view.

The Nelson Tunnel was driven during the eight-year period from April 1892 to June 1900. When completed, it drained many of the flooded portions of the mines along the Amethyst vein. The operating company obtained contracts to transport ore from these mines through the tunnel and along the connected surface rail tramway to the Humphreys mill at the Willow Creek forks. The ore was transported at one dollar per ton.

In 1891, Theodore Rennica staked the Last Chance on the Amethyst vein along West Willow Creek. The ore looked so good that Nicholas Creede staked the adjacent vein section as the Amethyst claim. Both mines prospered. In this view from 1895, the Amethyst is in the foreground with the Last Chance behind.

From 1902 to 1918, the Humphreys mill processed and concentrated ore hauled through the Nelson Tunnel. The mill was also partially powered by water wheels: water taken from the tunnel and delivered by flume fell through a 212-foot-long, 16-inch-diameter pipe to two Pelton water wheels, which were housed in a small building at the bottom of the mill. The Pelton wheels turned a 75-foot rope drive that powered the mill's drive shaft. In this photo, the 212-foot-long pipe and wheel pit building are clearly visible just to the left of the mill building.

Several Denver and Rio Grande gondola cars are waiting to be filled with concentrates from the Humphreys mill ore bins.

Bob Womack began prospecting along Cripple Creek in 1878, staked a claim in 1886, and in 1890, relocated his claim, naming it the El Paso. Assays reported 12 troy ounces in gold per ton, but there was no visible gold, just some silvery and silvery-yellow minerals that no one could identify. Ed De LaVergne identified the mineral as a gold-bearing telluride mineral—calaverite. In this scene from Cripple Creek's early days, the photographer has captured teamsters with their ore wagons at the Burns (or No. 1) shaft of Colorado's greatest gold mine, the Portland. By 1900, the Portland had produced more than $10,000,000 and paid $3,300,000 in dividends.

Using a measuring tape, Jimmy Burns and James Doyle located open ground in between some recently relocated claims on Battle Mountain above Victor. They found a triangular piece of ground, measuring 69/1,000 of an acre, that no one owned. There, they sank a 30-foot shaft and staked a claim—but they couldn't figure out whether they had any ore. For a one-third interest in the claim, John Harnan helped them identify the ore. Then, Winfield Scott Stratton took a quarter interest for financing the defense against adverse lawsuits. Those two trades—stocks for identification services and legal fees—were perhaps the wisest things that either Burns or Doyle ever did. This is a view of the Burns Shaft from around 1896.

Winfield Scott Stratton, a carpenter by training and a shrewd prospector through the school of hard knocks, set out to search for the gold-bearing telluride minerals in the spring of 1891. He found the contact of the Cripple Creek volcano with the surrounding granite and located the Independence claim on July 4, 1891. Eventually he came to own more productive mining interests in the district than anyone else. He sold the Independence to the London Venture Corporation for $10,000,000.

On April 25, 1896, a fire broke out in Cripple Creek and burned 30 acres. Four days later, another fire broke out that wiped out most of the town. Mine owners hired trains, filled them with supplies, and rushed them to the town. Cripple Creek rebuilt with brick and stone, creating the town visitors see today. In the center of this view, the large plume of smoke is the explosion of 700 pounds of dynamite at the Harder Grocery.

In August 1899, a fire started in a Victor dance hall and destroyed most of the business section of town. In this view, the Woods Brothers' Gold Coin shaft house is going up in flames. The Gold Coin shaft house and surface plant was rebuilt in brick—with stained-glass windows!

In this photograph, the Ajax mine crew pose in front of the shaft and headframe. Prominent in this view are the double-deck cages, having twice the capacity for lifting or lowering men, ore, supplies, and equipment. With a total production of more than $20,000,000, the Ajax was the sixth-most-productive mine in the District. It is located on Battle Mountain above Victor.

By 1900, the district's population exceeded 50,000. In addition to Cripple Creek and Victor, there were many other smaller towns scattered about—Goldfield, Independence, Altman, Hollywood, Cameron, Anaconda, Elkton, Gillett, and Midway. This photograph shows much of Anaconda laid out along a single street. The extensive cribbing around the Mary McKinney mine dumps is prominent at center right. Just behind the Mary McKinney is the Morning Glory, and further uphill is the Doctor-Jackpot Consolidation.

The Cresson mine was the second-most-productive mine in the district. In November 1914, miners on the 12th level broke into a 23-foot by 13-and-a-half-foot by 40-foot-high vug (an open space in the rock) that was lined with calaverite crystals. Fourteen hundred sacks of crystals were scraped from the walls, and 1,000 more sacks of lower-grade ore were chipped out before the rib, back, and floor of the vug were mined out. The Cresson vug produced nearly 60,000 troy ounces of gold worth some $1,200,000. Copies of the $468,637.29 check from Copeland Ore Sampling Company for purchasing the sacked ore were sold as souvenirs in Colorado for years.

Sampling works became an important business in Colorado's mining regions. Mining companies didn't trust smelters, and vice versa. So mining companies began to use third parties to test and assay their ore to determine a fair price before it was shipped to a smelter. Smelters, however, paid much better prices for larger lots of ore, so a sampling works could buy many smaller lots and then combine them into one larger lot that would bring a better price. By acting in this capacity as a middleman, the sampling works could pay a better price for small lots of ore from smaller mines. The Eagle Sampler at Goldfield was one of the most successful in the district.

By 1892, there was a Cripple Creek stock exchange. In spite of a strike in 1894, as well as the failure of several mines to pay dividends in 1895, three stock exchanges began operating on Cripple Creek's Bennett Avenue. A year later, 70 brokers were promoting the district's mines. Stock sales and mine production peaked in 1901 to 1902. As shown in this photograph at one of the district's stock exchanges, Spencer Penrose, co-owner of the C.O.D. (Cash on Delivery) mine with Charlie Tutt, seems quite content with his latest deal. Penrose later transformed a small resort into the famous Broadmoor Hotel in Colorado Springs.

The Vindicator mine at Independence was the fourth-leading producer in the District. It was also the scene of a brazen act of murder during the violent miners' strike of 1903–1904. Harry Orchard of the Western Federation of Miners labor union planted a dynamite bomb on the sixth level of the Vindicator that would explode when the lift cage was opened. He actually meant to set it on the seventh level, where he would have killed 15 scab workers at the shift change. Instead, he killed the mine superintendent and shift boss.

In 1904, Orchard dynamited the train station at Independence, killing 13 men outright and maiming another 14. Six survivors required amputations of arms or legs. Orchard's bombings helped turn the people against the union.

Governor James Peabody essentially declared war on militant unionism. As seen in this photo, the Colorado Militia established a threatening presence in downtown Cripple Creek with mounted cavalry and a Gatling gun.

The tents of the Colorado Militia were pitched near Stratton's Independence on Battle Mountain. The site soon became known as Camp Goldfield.

At Crapper Jack's taxi dance hall in Cripple Creek, a man could buy a dance with a girl for 25 cents. A live band played music, and the women made a commission on each dance purchased. Most of the women were not prostitutes.

The boys have brought in a friend for a beer at the White House Saloon.

MINING FOR INDUSTRY

(1893–1920)

New technologies that arose between 1860 and 1920 required metals in ever-increasing amounts. Industrial demand for metals shot up with the start of a European arms race in the 1890s. Key to this market was machining steel to manufacture everything from gun barrels to engine parts. The tools used to shape metal on high-speed lathes and mills needed to be made from metals that were harder than those being machined. Tungsten-steel alloys provided that hardness and strength. The Boulder County Tungsten District soon became the world's leading producer of this metal.

Both vanadium and molybdenum are also useful for hardening steel. The use of vanadium as an alloy compound was discovered at about the same time as tungsten. Vanadium alloys are also very corrosion-resistant. Many uranium-vanadium deposits on the Colorado Plateau were mined during the early years of the twentieth century. Metallurgists were much slower to develop molybdenum-hardened steel alloys, so the world's largest molybdenum deposit at Climax did not become truly important until after the end of World War I.

Various lead minerals are ubiquitous in most silver deposits. Lead for solder, plumbing, paint, batteries, and bullets provided another boost to the badly sagging silver mines. In addition to lead, several zinc minerals are abundant in silver deposits. Zinc is the key component for the galvanized coating that can be applied to iron and steel to prevent rust, and it is also the primary metal mixed with copper to create the alloy brass. With the decline in silver prices, lead and zinc production became important at Leadville, Gilman, Aspen, Creede, Telluride, Silverton, and Silver Plume.

Copper became important in all sorts of electrical applications. Although Colorado is not known for its copper mineralization, both Gilman and Leadville produced a fair amount as a by-product of their overall metal mining efforts.

Colorado Fuel and Iron Company operated several iron mines in the state. The most important were the Orient mine in Saguache County, and the Calumet mine in Chaffee County. Key to the steel-making process, fluorspar was mined at Wagon Wheel Gap in Mineral County. Today, Colorado continues to mine minerals for use in industry.

Following the 1893 Silver Panic, mines were faced with a stark economic reality—they could mine only the part of the deposit that was richest in silver, mine something else that was profitable, or take measures that would further the economy of the mining operation. In Leadville, the A. Y. and Minnie, Colonel Sellers, and Moyer began mining high-grade lead and zinc ore with silver. Silver became a bonus to base-metal production.

Mining man Samuel D. Nicholson discovered large bodies of oxidized zinc ore, amenable to new metallurgical processes, in many of the older silver mines on Carbonate Hill in Leadville. He acquired leases and operating agreements and began mining. At this time, the invention of galvanization—applying zinc coatings to other metals to prevent rust—provided a tremendous boost to the zinc market. This photo shows the flagship of Nicholson's zinc mining operations, the Wolftone mine.

Opportunities to mine minerals for industry attracted the attention of politicians. To celebrate zinc, Nicholson invited everyone who was anyone in Colorado to an underground banquet at the Wolftone on January 25, 1911. This photo shows the serving crew awaiting the guests' arrival.

In the late nineteenth century, it was discovered that steel alloyed with tungsten was significantly harder and tougher than steel alone. People quickly discovered that tungsten-steel tools did a much better job of cutting and shaping metal objects on high-speed lathes and mills. In 1899, Samuel Conger and W. H. Wanamaker found tungsten deposits near Nederland, Boulder County. During that time, the arms race among European powers provided a market for manufacturing tungsten-steel guns and armaments. Suddenly, Boulder County became the world's leader in tungsten production. The Vasco Number 6 mine, shown in this photograph, was one of the county's leading producers.

The Wolf Tongue Mining Company acquired the old Caribou mill in Nederland, refurbished and re-equipped it to separate tungsten, and began processing its own ore. Wolf Tongue is a play on the German word for tungsten, *wolfram,* combined with the English word *tongue* for tungsten.

Conger and Wanamaker's discovery, the Conger mine, was purchased by the Primos Company, which built this mill at Lakewood, just north of Nederland. In 1906, the Primos was the world's largest tungsten production facility. Lakewood no longer exists, and the ruins of the Primos mill consist of a few foundations on the hillside.

Vanadium was another element discovered that hardens steel in alloys. Vast deposits of vanadium and uranium were found on the Colorado Plateau in western Colorado and eastern Utah. This scene was photographed in a stope at one of the Primos Company's vanadium mines. The candle lighting indicates that the photograph was probably taken before 1906.

This photograph shows the mill of the United States Vanadium Mining Company at Rifle, a town in western Colorado's Garfield County.

This photograph shows the Wagon Wheel Gap fluorspar mine and mill and the Hot Springs Resort in Mineral County. Before World War I, the mineral fluorite, or fluorspar, was used as a flux in melting iron ore in the steel-making process. It was also used to manufacture hydrofluoric acid for the chemical industry. The most important use for fluorspar and hydrofluoric acid—producing aluminum—arrived in the 1920s, when demand for aluminum began to grow significantly. Hydrofluoric acid, sodium oxide, and aluminum oxide are used to produce artificial cryolite, which allows aluminum to be separated from bauxite in a furnace. This method (named the Hall-Heroult process for the discoverers) is the only production method for aluminum that has been used since 1886.

Two iron deposits were mined in Colorado by the Colorado Fuel and Iron Company. This view of Orient in Saguache County, location of the Orient Iron mine, shows several bunkhouses and a boardinghouse.

In Eagle County, above the Eagle River Canyon, the town of Gilman is visible as a line of buildings on the hill. During the nineteenth century, the Gilman District was home to several small silver and gold mining operations on the cliffs below town. At lower right, a mine railway trestle and the roofs of several mine buildings can also be seen.

At Belden, on the floor of the Eagle River Canyon, a mill was built to process ore from the mines above, located along the cliffs. This view shows the Belden inclined rail tramway and the milling facility. The Denver and Rio Grande Railroad route along the Eagle River provided shipping for the mill concentrates.

New Jersey Zinc Company began acquiring old mining claims after its studies showed significant zinc ore in the Gilman ore deposit. The company named its consolidation of claims the Eagle mine, and they sank the Wilkesbarre shaft (shown in this photograph) to mine the ore body. The Eagle mine operated into the 1980s and became Colorado's greatest zinc producer. It was also the state's most important producer of mineral specimens.

In the Fremont Pass area, prospectors discovered significant deposits of a mineral often cursed by earlier-day prospectors looking for gold and silver: "molly-be damned," or molybdenite, as it is known today. Eventually, it was found to be useful in hardening steel, and a growing alloy industry slowly began to demand more of the metal. This photo shows an infant mining operation called Climax on Bartlett Mountain. Climax conducted a far-ranging research program aimed at discovering uses for molybdenum, the element contained in the mineral molybdenite. In the process, Climax became the world's leading producer of the metal.

Notes on the Photographs

These notes, listed by page number, attempt to include all aspects known of the photographs. Each of the photographs is identified by the page number, photograph's title or description, photographer and collection, archive, and call or box number when applicable. Although every attempt was made to collect available data, in some cases complete data was unavailable due to the age and condition of some of the photographs and records.

HISTORIC PHOTOS OF
COLORADO MINING

In 1859, 100,000 folks started the journey to the Pikes Peak goldfields, but only 50,000 completed the trip. An additional 25,000 soon gave up and went back home. The remainder not only brought statehood to the central Rocky Mountains, but they also brought the industrial world to isolated areas in the high mountains, where they mined mineral deposits for gold, silver, lead, zinc, and copper, among others.

This book, *Historic Photos of Colorado Mining*, provides an introduction to Colorado's mining history through photographs from the nineteenth and early twentieth centuries. Accompanying captions provide specific contexts for the photos and tell the story of the prospectors, miners, engineers, teamsters, railroaders, and townspeople who served as entrepreneurs and workers in industrializing the Colorado Rocky Mountains.

Many ruins from the mining days are now recognized as historic landmarks. But the stories behind the ruins are often as fascinating as the ruins themselves—the struggle to survive and thrive in the wilderness is always a compelling tale.

Ed Raines is a consulting geologist, mineralogist, and mining historian. For the past year and a half, he has served as consulting Collections Manager for the Geology Museum at the Colorado School of Mines. Ed has written numerous papers on the mining history, geology, and mineralogy of various mining districts and mineral deposits. Three times he has been recognized by the Friends of Mineralogy for Author of the Year's Outstanding Paper in the journal *Rocks and Minerals*. He has coordinated conferences for the Mining History Association, leading field trips and presenting papers at each conference at Leadville, Creede, and Gold Hill.

Ed served on the Boulder County Historic Preservation Advisory Board for nine years and chaired the board for two years. In 2000, he received an award for his historic preservation work at Leadville. Prior to 1988, he worked as a petroleum geologist.

WWW.TURNERPUBLISHING.COM

www.ingramcontent.com/pod-product-compliance
Lightning Source LLC
Chambersburg PA
CBHW061227150426
42812CB00054BA/2538